Quantum for Beginners

A Guide to Help you learn about how Universe Began, About Pioneers of Physics, Famous Experiments, Black Holes, and Quantum Field Theory

By

ADELE FOSTER

© Copyright 2022 by ADELE FOSTER- All rights reserved.

This document is geared towards providing exact and reliable information in regard to the topic and issue covered. The publication is sold with the idea that the publisher is not required to render accounting, officially permitted, or otherwise, qualified services. If advice is necessary, legal, or professional, a practiced individual in the profession should be ordered.

- From a Declaration of Principles, which was accepted and approved equally by a Committee of the American Bar Association and a Committee of Publishers and Associations. In no way is it legal to reproduce, duplicate, or transmit any part of this document in either electronic means or in printed format. Recording of this publication is strictly prohibited, and any storage of this document is not allowed unless with written permission from the publisher. All rights reserved. The information provided herein is stated to be truthful and consistent, in that any liability, in terms of inattention or otherwise, by any usage or abuse of any policies, processes, or directions contained within is the solitary and utter responsibility of the recipient reader. Under no circumstances will any legal responsibility or blame be held against the publisher for any reparation, damages, or monetary loss due to the information herein, either directly or indirectly. Respective authors own all copyrights not held by the publisher. The information herein is offered for informational purposes solely and is universal as so. The presentation of the information is without contract or any type of guarantee assurance. The trademarks that are used are without any consent, and the publication of the trademark is without permission or backing by the trademark owner. All trademarks and brands within this book are for clarifying purposes only and are owned by the owners themselves, not affiliated with this document.

Hi, my name is Adele and I am passionate about physics and astronomy. When I was studying at college.

I used to entertain my nephews and their friends with science fiction stories that I invented on the spot, but based on real physics studies. The twin paradox was their favorite, they always asked me!

It was easy for me to decide what to study, so I enrolled in theoretical physics at Yale, where I am now completing a PhD in nuclear astrophysics. They are very complex subjects, I admit, but I am sure that with the right words everyone can understand the extent and influence that quantum physics has in our lives, even if we do not always realize it.

I have written this book for all those who would like to know more, but do not have a solid basis in the subject nor too much time to devote to it. There will be no complicated mathematical formulas and the concepts will be explained in the simplest way

I have found. I have made a great effort to make this book simple and engaging, but without losing scientific rigour. If you search on Amazon, you will see that many other books in this genre talk about the law of attraction. But pay attention! It's just a commercial stratagem to sell more books, the law of attraction has nothing to do with quantum physics and has no scientific basis. Here you will only read what is scientifically proven, no small talk.

If you find this book helpful, I would appreciate it if you left an honest Amazon review!

And now ... let the journey begin!

Table of Contents

Introduction 7

Chapter 1: Origin Of Universe 11

1.1 Greatest Minds 11

1.2 Origin Of Universe 16

1.3 The Age Of The Solar System 22

1.4 Quantum Theory Of Light 24

1.5 Photons 30

Chapter 2 Quantum Technology 37

2.1 What Is Quantum Technology 37

2.2 History Of Quantum Technology 39

2.3 Phenomena Of Quantum Technology 41

2.4 Four Areas Of Quantum Technology 44

2.5 Myths About Quantum Physics 47

2.6 Difference Between Quantum And Classical Physics ... 53

Chapter 3 Quantum Physics And Experiments 57

3.1 Rutherford Experiment 57

3.2 Quantum Tunneling 61

3.3 Schrodinger's Cat 64

3.4 Quantum Field Theory 67

3.5 Quantum Gravity *69*

3.6 Quantum Entanglement *73*

Chapter 4: Black Holes **77**

4.1 Black Hole *77*

4.2 Hawking Radiation *80*

4.3 What Is The Size Of Black Holes? *82*

4.4 Stellar Black Holes *83*

4.5 Intermediate Black Holes *83*

4.6 The Creation Of Giants *84*

4.7 What Causes Black Holes To Form? *86*

4.8 How Do Scientists Know There Are Black Holes If They Are "Black"? *86*

4.9 What Do Black Holes Look Like? *87*

4.10 Binary Black Holes Are Being Illuminated. *88*

4.11 Weird Facts About Black Holes *89*

Chapter 5 Quantum Physics And Daily Life **99**

Conclusion **107**

Introduction

The study of quantum physics was founded in the late 1800s due to a series of experimental discoveries about atoms that did not make intuitive sense within the constraints of classical physics. It was one of the most basic discoveries when it was realized that matter and energy might be thought of as discrete packets, or quanta, each of which has a corresponding minimum value associated with it. For example, light with a certain frequency will supply energy in the form of quanta known as "photons." This frequency produces photons with the same energy as the previous one, and this radiation cannot be broken into smaller bits. "Quantum" derives from the Latin language and translates as "how much."

The study of quantum principles has completely transformed our view of the atom, which consists of a nucleus surrounded by electrons and the universe in general. The electrons in early models were portrayed as particles that orbited the nucleus in the same way that satellites orbited the Earth. At the moment, electrons are thought of in current quantum physics as being scattered within orbitals, which are mathematical representations that reflect the chance of an electron being in more than one place within a specific range at any given time. Because of this, electrons may jump from one orbital to another when they gain or lose energy, but they cannot travel across orbitals.

Quantum physics describes how atoms function and how chemistry and existence function as a whole. In a way, you, I, and the gatepost are all dancing to the same melody, at least on some level. How electrons go through a computer chip, how photons of light are turned into electrical current in a solar panel, how laser photons duplicate multiplied, and how the sun continues to burn are all explained by quantum physics.

This is the point at which the complexity – and the fun for physicists – really starts. To begin, there is no such thing as a single quantum theory. There's quantum mechanics, which is the underlying mathematical underpinning that underpins everything, developed in the 1920's by Bohr, Heisenberg, Schrödinger, and other researchers. Basic phenomena such as the change in the position or dynamism of a single particle or a collection of particles over time are described by quantum mechanics.

It is necessary to link quantum mechanics with other components of physics, another very remarkably Albert Einstein's special theory of relativity, which describes what happens when objects move extremely quickly, to construct quantum field theories, which explain how things function in the real world.

Three distinct quantum field theories are discussed three of the four fundamental forces of nature that interact with substance:

- Electromagnetism explains how atoms hold together.

- The strong nuclear force explains why some atoms decay radioactively.
- Indeed, the weak force discusses why some atoms decay.

Electromagnetism is one of the phenomena interactions that interact with matter. Physicists refer to the study of matter and energy as quantum physics at its most basic level. Its purpose is to understand better the features and behaviors of the basic building blocks of nature.

While many quantum experiments are conducted on tiny objects which include electrons and photons, quantum phenomena may be seen in a wide range of sizes and are all around us. However, when it comes to larger issues, we may not detect them as soon. It can generate the impression that quantum processes are odd or out of the ordinary. Indeed, quantum inquiry fills in the gaps in our knowledge of physics, providing us with a vision of the world in which we live daily.

Our basic understanding of substances, chemistry, biology, and astronomy has been enhanced due to quantum discoveries. These discoveries have helped in the progress of technologies like lasers and transistors and the advancement of technologies that were previously supposed to be wholly speculative, such as quantum computers.

Theoretical physicists investigate whether quantum physics will alter our understanding of gravity and its link to space and time. Quantum research may reveal how everything about the universe (or multiple universes) is connected to everything else at higher levels of abstraction that our faculties cannot see.

Chapter 1: Origin of Universe

1.1 Greatest Minds

1. Albert Einstein

Albert Einstein is regarded as one of the twentieth century's most significant scientists. His work continues to aid astronomers in their research into anything from gravitational waves to the orbit of Mercury.

Even individuals who don't grasp the underlying physics of $E = mc^2$ – the equation that explain special relativity – are familiar with it. Einstein is also recognized for his general theory of relativity (which explains gravity) and the photoelectric effect, for which he won the Nobel Prize in Physics in 1921. Einstein also endeavored in vain to unite all of the universe's forces in a single theory, or theory of everything, which he was still working on when he died.

2. **Niels Bohr**

Niels Bohr was one of the most significant scientists of the twentieth century, well known for his contributions to quantum theory and Nobel Prize-winning studies on the structure of atoms.

Bohr was born in Copenhagen in 1885 to well-educated parents and had an early interest in physics. In 1911, he obtained a Ph.D.

in physics from Copenhagen University after studying the topic throughout his undergraduate and graduate years.

Bohr won a competition sponsored by the Academy of Sciences in Copenhagen while still a student for his research on measuring liquid surface tension using oscillating fluid jets. Bohr did various tests and even constructed his glass test tubes while working in his father's (a famous physiologist) laboratory.

3. **Erwin Schrödinger**

Erwin Schrödinger was an Austrian theoretical physicist who pioneered the wave theory of matter and other principles of quantum mechanics. On August 12, 1887, he was born in Vienna, Austria, and died in Vienna on January 4, 1961. He and

British scientist P.A.M. Dirac received the Nobel Prize in Physics in 1933.

Schrödinger proposed a hypothesis that uses a wave equation to describe the behavior of such a system, which is now known as the Schrödinger equation. Unlike Newton's equations, the solutions of Schrödinger's equation are wave functions that can only be attributed to the likelihood of physical events occurring. In quantum mechanics, the definite and easily seen sequence of events of Newton's planetary orbits is replaced with the more abstract concept of probability.

4. Werner Heisenberg

Werner Heisenberg, full name Werner Karl Heisenberg, was a German physicist and philosopher who developed (1925) a technique to define quantum physics in terms of matrices. On December 5, 1901, Heisenberg was born in Wurzburg, Germany, and died on February 1, 1976, in Munich, West Germany. He was granted the biggest prize, Nobel Prize in Physics, in 1932 for his discovery. He introduced his uncertainty principle in 1927, which functioned as the basis for his philosophy and for which he is most known. He also made significant contributions to theories of turbulent flow hydrodynamics, atomic nucleus, cosmic rays, and particles, and he played a key role in the planning of the West German nuclear reactor in Karlsruhe, as well as a research reactor in Munich, in 1957. His work on atomic study during World War II has sparked a lot of debate.

5. **Max Planck**

Max Planck (born April 23, 1858, Kiel, Schleswig [Germany]—died October 4, 1947, Gottingen, Germany) was a German theoretical physicist who invented quantum theory; he received the Nobel Prize in Physics in 1918.

Planck made several contributions to theoretical physics, but he is most known for creating quantum theory. Just as Albert Einstein's theory of relativity altered our knowledge of space and time, this theory revolutionized our knowledge of atomic and subatomic processes. They are the foundational ideas of twentieth-century physics when taken together. Both have compelled humanity to rethink some of its most deeply held philosophical assumptions, and both have resulted in industrial and military applications that touch every facet of contemporary life.

1.2 Origin of Universe

- **The Big Bang**

Since the early 1900s, one hypothesis for the universe's origin and destiny, the Big Bang theory, has dominated the debate. According to the Big Bang theory proponents, all of the matter and energy in the known universe was crowded into a small, compact point between 13 billion and 15 billion years ago. In fact, according to this hypothesis, matter and energy were the same at the time, and it was impossible to tell them apart.

The Big Bang theorists think a tiny but dense primal matter/energy point burst. The fireball expelled matter/energy at speeds nearing the speed of light in a couple of seconds. Energy and matter started to break away and form independent things later—maybe seconds later, perhaps years later. What poured out of this first explosion gave rise to all of the various elements in the cosmos today.

According to Big Bang theorists, all galaxies, stars, and planets still have the explosive momentum of the instant of creation and are rapidly racing away from one other. This hypothesis was based on an extraordinary discovery concerning our nearby galaxies. In 1929, astronomer Edwin Hubble of the Mount Wilson Observatory in California stated that all of the galaxies he had discovered were retreating from us and each other at rates of up to thousands of miles per second.

- **The Redshift**

Hubble used the Doppler Effect to measure the speeds of these galaxies. When a source of waves, such as light or sound, moves concerning an observer or listener, this phenomenon happens. When an origin of sound or light moves toward you, the waves seem to rise in frequency: music grows higher in pitch, and light shifts toward the blue end of the visible spectrum. The frequency of the waves falls as the source moves away from you: sound gets lower in pitch, and light tends to shift toward the red end of the spectrum. When you listen to an ambulance siren, you may detect the Doppler Effect: the sound rises in pitch as the vehicle approaches and lowers in pitch as the vehicle speeds away.

Hubble utilized a spectroscope, an instrument that studies the multiple frequencies present in light, to investigate the light from the galaxies. He detected that the light from distant galaxies was pushed toward the red end of the spectrum. It didn't matter where each galaxy was in the sky; they were all redshifted. Hubble attributed the change to the galaxies being in motion, zipping away from Earth. Hubble thought that the higher the redshift, the faster the galaxy.

There was just a small redshift in certain galaxies. On the other hand, others pushed their light beyond red and into the infrared, even down to microwaves. The redshifts of fainter, more distant galaxies seemed to be the greatest, indicating that they were moving the quickest of all.

Astronomical studies over the last century have shown that we live in an expanding universe that began in an extraordinarily dense and hot starting state termed the Big Bang a long but finite time ago. Aside from matter, space is an important component of our cosmos, and the observable expansion of the universe means that the quantity of space available grows with time. There was less room in the past, and there will be more in the future. It seems bizarre and counterintuitive to our everyday experience, in which we perceive space as a fixed amount, such as the volume of our home. However, physics informs us that space, even if it is a pure vacuum (space with no atoms or molecules), is a vital component of the cosmos, containing hidden particles and energy and having the ability to expand and shrink.

The cosmos space holds the majority (about 70%) of all energy in the universe, according to a remarkable conclusion discovered in 1998 based on observations of the brightness of extremely distant exploding stars.

This energy is manifested as a mysterious force that accelerates the expansion of the cosmos, which is currently unknown.

The remaining about 30% of the universe's energy (Einstein says mass and energy are identical) presents itself as the mass of "actual" matter, which attracts each other gravitationally. Only roughly a sixth of this actual stuff is regular matter, consisting of atoms and molecules. The other half is enigmatic Dark Matter, which exerts gravitational pull but whose nature is unknown.

Albert Einstein predicted the presence of "anti-gravity" in 1916. He had no opinion at the time that this would result in a force that would speed up the expansion of the cosmos. Willem de Sitter, a Dutch astronomer, found this in 1917. Einstein later referred to the concept of anti-gravity as his "biggest error." It wasn't until 1998 that it was realized that this was not a mistake but rather a brilliant revelation.

We also find the concept of time having a beginning puzzling. We are used to the idea that a street has a beginning and an end, but we get the impression that time is flowing ahead inevitably, from an eternally distant past to an indefinitely distant future. Nonetheless, many astronomical studies have shown that time began around 13.8 billion years ago. There was no time before then.

All of this looks weird and contradictory to our instincts and emotions. However, many physical events and rules seem to contradict our everyday experience and intuition if one thinks about it. We no longer consider these facts and rules "abnormal" after being used to them. When we gaze out the window, for example, the Earth seems flat and at rest. But we know it's a sphere, that individuals in New Zealand are walking backward when seen from the Northern Hemisphere, and that the Earth revolves around its axis in roughly 24 hours. The speed of rotation near the equator is around 0.5 km/s, or 1800 km/h, double the speed of a commercial jet airliner. The rotational speed is still about 1300 km/h in New York's geographical latitude. We also know that the Earth travels around the sun at a speed of around 30 kilometers per second, or 108,000 kilometers per hour. In roughly three and a half hours, we'll have covered the distance between here and the Moon.

But we sit calmly in our chairs, oblivious that we are traveling at such high rates. We have the impression that we are entirely at ease. Galilei Galileo (1564–1642), an Italian scientist and astronomer, was the first to observe that we are traveling at a constant pace. He spotted men in Venice on a passing ship tossing cargo bags to each other, and he noted that the sacks traveled between them as though the guys were resting and throwing the sacks to each other. He realized that everything we do (and the physics rules) continues as if we are at rest as long as we travel constantly. If we are in a car without windows, we have no means of knowing whether the vehicle is standing or traveling at a constant pace. A person traveling at a constant pace sees all physical events in the same manner that he or she does while at rest (insofar as it is possible to define "being at rest" in an absolute sense, which is not the case). Everyone who has flown knows this: the aircraft soars high in the sky at a constant speed of about 900 km/h, yet we stroll down the aisle, sip our coffee, and do everything we typically do at home on the ground. Galilei observed that the only change in speed that can be seen is an acceleration or slowdown (a negative acceleration).

We are forced into the backs of our seats as a train or aircraft accelerates, and we rocket forward when the vehicle abruptly stops. That is why we must fasten our seatbelts. The law of inertia states that an object prefers to maintain its current velocity and "opposes" a change in velocity.

1.3 The Age of the Solar System

It is stated numerous times in the preceding that the solar system is roughly 4.6 billion years old. It is the age of the oldest meteorites discovered on the planet.

When a tiny stone or rock from space reaches Earth's atmosphere, a meteor is a flaming trail that appears in the sky for a short while. Earth travels around the sun at a speed of around 30 km/s, whereas objects are circling the sun in our neighborhood travel at speeds of tens of kilometers per second. When such a stone enters Earth's atmosphere, it travels much faster than the Space Shuttle or astronauts' space capsules returning from the International Space Station, which travel at a speed of 8 km/s and have a heat shield that glows hot due to friction caused by the stone's entrance velocity of 28 times the sound velocity. Similarly, due to the friction between the stone and the air, the stone becomes incredibly heated and starts to shine and evaporate. The hot vapors and heated air shine brilliantly, creating a flaming trail across the sky known as a meteor. If a chunk of rock remains after passing the atmosphere and is discovered on the ground, it is a meteorite. Small stones, just a few inches in diameter, fully burn up.

The ensuing meteor dust gradually falls to Earth's surface, where it may be found in the ice caps of Greenland and Antarctica, as well as on the ocean bottoms. A stone from space must weigh at least a few pounds or, by chance, have an extremely low velocity about Earth as it reaches the atmosphere to reach the ground.

As described in the box, the age of rocks, including meteorites and the Earth's crust, is established by analyzing the decay of radioactive elements contained in the rock. According to this study, the oldest meteorites and lunar rocks delivered to Earth by the Apollo astronauts have ages ranging from 4.60 to 4.65 billion years.

Because no earlier rocks have been discovered in the solar system, it is currently widely considered that the solar system originated 4.65 billion years ago. The universe is about 13.8 billion years old, and our Milky Way galaxy existed at least 13 billion years ago. It means that when the sun and its family of planets were formed in an interstellar cloud in the corner of our Milky Way, the cosmos had already existed for nine billion years. The Earth's crust and the black basalt planes on the Moon are less than 4.6 billion years old. There have been no rocks discovered on Earth older than 4.3 billion years. The Earth's crust was hot and liquid initially, with the first crustal rocks solidifying just 4.3 billion years ago. These oldest rocks may be discovered in Hudson Bay in Canada.

1.4 Quantum Theory of Light

- **Wave-Particle Duality of Light**

According to quantum theory, both light and matter are made up of microscopic particles with wavelike qualities. Photons are the particles that make light, whereas electrons, protons, and neutrons are the particles that make up matter. The wavelike qualities appear only when its mass is tiny enough. Let's look at how light acts as a wave and as a particle to better grasp everything.

- **Wavelike Behavior of Light**

A Dutch scientist named Christiaan Huygens demonstrated that light behaved like a wave in the 1600s.

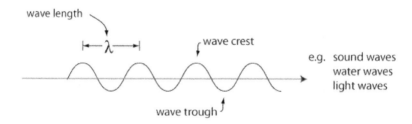

Diffraction is one of the wave's properties.

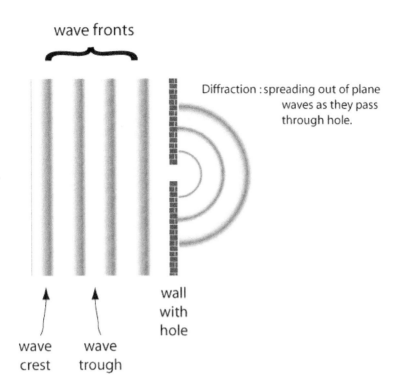

The wave diffracts less when the slit width becomes greater than the wavelength.

Interference is another wave behavior.

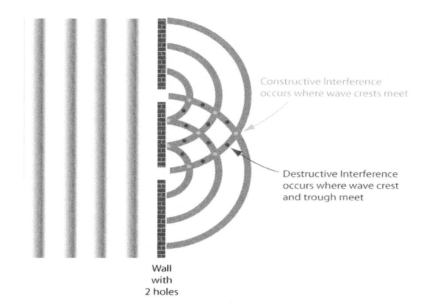

Constructive Interference occurs where wave crests meet

Destructive Interference occurs where wave crest and trough meet

Wall with 2 holes

In the 1800s, James Clerk Maxwell demonstrated that light is an electromagnetic wave that moves at the speed of light across space. According to the law of energy conservation, the frequency of light is proportional to its wavelength.

$$\text{frequency} \rightarrow \nu = \frac{c \leftarrow \text{speed of light}}{\lambda \leftarrow \text{wavelength}}$$

Let's have a look at a hypothetical computation.

The light blue glow given off by mercury street lamps has a wavelength of $\lambda = 436$nm. What is its frequency?

$$\nu = \frac{c}{\lambda} = \frac{3.00 \times 10^8 \text{ m}}{\text{s} \mid 436\text{nm}} \mid \frac{10^9 \text{ nm}}{\text{m}} = 6.88 \times 10^{14} \text{ s}^{-1}$$

The unit s^{-1} is so common when talking about waves that it was given the name Hertz. That is, $1 s^{-1} = 1$ Hz. Thus, we would say that light with a wavelength of 436 nm corresponds to a frequency of 6.88×10^{14} Hertz.

The visible area of electromagnetic radiation is defined as the range of wavelengths between 400 and 750 nm that can be seen with the naked eye.

As we saw in the sample above, blue light is close to our eyes' high-frequency limit. The low-frequency limit of human eyes is a red light with wavelengths about 750 nm.

White light is light that comprises all frequencies observable in the visual spectrum.

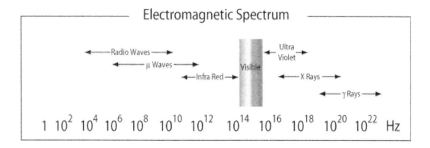

Various names are given to different areas of the electromagnetic spectrum in general. The names given to light's distinct regions (frequency ranges) based on their frequency range are listed below.

- **Light Has Particle-Like Behavior**

You could believe that light acts like a wave are self-evident. So, how do we know that photons are the particles that makeup light? An experiment known as the photoelectric effect lends credence to this theory.

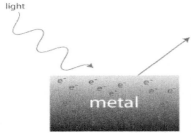

Light is shown on a metal and after a certain binding energy is overcome, an electron is emitted from the metal.

The electron is expelled from the metal with specified kinetic energy, a crucial component of this experiment (i.e., a specific speed). Anyone who understands how waves behave knows that the energy associated with a wave is proportional to its amplitude or intensity. For example, in the ocean, the larger the wave, the greater the energy connected with it. It's the huge waves that knock you over, not the tiny ones! When the intensity of the light was raised (brighter light), the kinetic energy of the released electron did not change; everyone who believed light was simply a wave was thrown for a loop. More electrons are released as the light becomes brighter, but they all have the same kinetic energy.

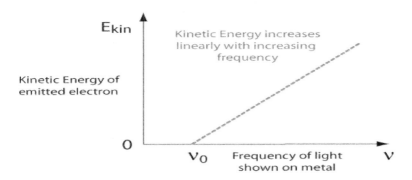

They reasoned that the kinetic energy of the released electron had to be affected by something. As a result, they modified the light's frequency, which affected the kinetic energy of the released electron. Each metal, however, has a threshold frequency, v0, below which no electrons are released. It means that kinetic energy equals the frequency of light multiplied by a constant (i.e., the slope of the line). The constant is Planck's constant and is denoted by the letter h.

$$h = 6.63 \times 10^{-34} \text{ J} \cdot \text{s} \leftarrow \text{Planck's Constant}$$

The kinetic energy of the expelled electron may now be written as an equation.

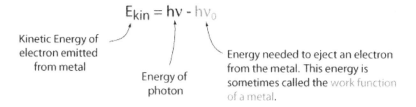

This discovery contradicts the traditional view of light as a wave. Light arrives in individual bundles called photons, and each photon must have enough energy to release a single electron, according to one theory that fits this image. Aside from that, nothing occurs. As a result, a single photon's energy is:

$$E_{photon} = h\nu$$

It was a shocking conclusion when it was initially realized. Albert Einstein was the first to state the photoelectric effect, for which he was awarded the Nobel Prize in Physics.

1.5 Photons

A photon is a very small particle made up of electromagnetic waves. Photons are just electric fields flowing across space, as shown by Maxwell. Photons are charged particles with no rest mass that move at the speed of light. Photons are released by charged particles colliding, although they may also be emitted by other means, such as radioactive decay. Because photons are such tiny particles, their wavelike features profoundly impact their behavior. A squiggly arrow in diagrams depicts individual photons.

- **Description**

Photons are often referred to as energy packets. It is an excellent comparison because a photon carries energy that cannot be split. An oscillating electric field is used to store this energy. Almost any frequency may be used to oscillate these fields. The largest theoretical wavelength of light is the size of the universe, while some theories anticipate the smallest conceivable at the Planck length, even though they have never been detected. These energy packets may be sent across long distances without energy loss or speed. In space, photons move at the speed of light, 2.997x10⁸ m/s. The speed of an electric field in space may be used to calculate the speed of a photon across space. In 1864, Maxwell revealed this evidence. Photons have a measurable momentum that follows the de Broglie equation, even with no mass. Photon momentum allows for novel practical applications such as optical tweezers.

Photons, in general, exhibit qualities that are comparable to electromagnetic waves. A wavelength and frequency are assigned to each photon. The wavelength is the distance between two electric field peaks with the same vector. The frequency of a photon is defined as the number of wavelengths it travels in one second.

A photon, unlike an electromagnetic wave, cannot have a color. Instead, a photon will correlate to a certain hue of light. Because the human eye's capabilities determine color, a single photon cannot have color because the human eye cannot perceive it. Several photons must interact with the retina to detect and record the light of a certain color.

Color can only be experienced when multiple photons operate in concert on the retina as an electromagnetic wave.

- **According to Maxwell's Equations**

Maxwell's equations are the most precise explanations we have of the nature of photons. Maxwell's equations describe how photons flow across space mathematically. An orthogonal magnetic field is created when an electric field undergoes flux.

The flow of the magnetic field then recreates the electric field. The wave pair may travel across space at the speed of light due to the formation and destruction of each associated wave. Within the context of quantum dynamics, Maxwell's equations accurately explain the nature of individual photons.

- **Photon Production**

Photons may be created in a variety of ways. Photons may be emitted in various ways, which will be discussed in this section. Photon emission necessitates the passage of charged particles since photons are electric fields propagating across space.

- **Blackbody Radiation**

The atoms inside a material vibrate at greater energies when it is heated. These oscillations quickly change the shape and energy of electron orbitals. Photons are released and absorbed at energies proportional to the energy of the change as the energy of the electrons changes. The heat of an item may be sensed from a vast distance because of blackbody radiation, which causes light bulbs to glow. The use of black bodies to simplify objects allows for indirect temperature computation of distant objects. Astronomers and kitchen infrared thermometers use this concept daily.

- **Spontaneous Emission**

When elections descend from an excited to a lower energy state, photons may be spontaneously emitted (usually the ground state). Relaxation is the precise word for this decrease in energy. Based on the available energy levels in their surroundings, electrons undertaking this sort of emission will emit a highly different set of photons. An emission spectrum is built on this collection of potential photons.

- **Fluorescence**

Florescence is a kind of spontaneous emission that is unique. The energy of a photon released inflorescence does not match the energy needed to excite the electron. When an electron loses a significant quantity of energy to its surroundings before relaxing, it will glow. Florescence is often used to observe the presence of target molecules in a laboratory environment. UV radiation is used to excite electrons, releasing visible light that researchers may see.

- **Emissions Stimulated**

A photon matching the difference between these energy levels may artificially lead an excited electron to relax to a lower energy state. The generated photon's phase and orientation and its energy and direction will be equal to that of the incoming photon's electric field. The light emitted by stimulated emission is considered coherent since it is identical to the photon that created it in every aspect. Lasers use stimulated emission to create coherent electromagnetic radiation.

- **Synchrotrons (electron bending)**

When the route of electrons with very high kinetic energy, such as those found in particle accelerators, is changed, high-energy photons are produced. A strong magnetic field is used to effect this change. It is true for all free electrons, but synchrotron radiation has unique practical consequences. Synchrotron radiation is presently the finest method for creating exact frequencies of directed x-ray radiation. Because of the high quality of x-rays generated, synchrotrons such as the Advanced Light Source (ALS) at Lawrence Berkeley Labs and the Stanford Synchrotron Radiation Light Source (SSRL) are hotbeds of x-ray spectroscopy.

- **Nuclear Decomposition**

High-energy photons may be released during some kinds of radioactive decay. Nuclear isomerization is one such form of decay. A nucleus rearranges itself to a more stable structure and releases a gamma-ray during isomerization. Proton decay will likewise release very high-energy photons, albeit this is merely a theory.

- **The Effect of Photo Electricity**

When light strikes a metal plate, electrons may break off from the plate's surface. The photoelectric effect is the result of this interaction between light and electrons. The photoelectric effect was the first clear proof that light beams were made up of quantized particles. The energy needed to evict an electron from a metal's surface is generally on the same order of magnitude as the energy necessary to ionize the metal. Because metals' ionization energies are typically many electron volts, the photoelectric effect is usually detected with visible or even higher energy light. The light was supposed to move in waves when this phenomenon was examined. Increasing the intensity of light resulted in a rise in current, not an increase in the kinetic energy of the released electron, as expected by the wave model of light. Later, Einstein addressed the discrepancy by demonstrating that light is made up of quantized packets of energy known as photons.

Because current may be created from a light source, the photoelectric effect has numerous practical uses. The photoelectric effect is often employed as a component in light-responsive switches. Nightlights and photomultipliers are two examples. In most cases, the current is so little that it has to be amplified before being used as a switch.

Chapter 2 Quantum Technology

Quantum technology is a sophisticated field of physics that focuses on the analysis of subatomic particles smaller than atoms, which are the core components of all matte and their interactions with one another.

2.1 What is Quantum Technology

When it comes to physical principles, quantum mechanics is the theoretical framework for understanding them at the atomic level. It first arose in the early 20th century, and it has since grown to become an essential aspect of the technological sphere, connected with significant and often confusing events. The capacity of a system to exist in numerous states simultaneously (superposition) or to display connectivity between particles despite great physical distances are examples of such capabilities (entanglement).

Numerous widely used technologies, such as transistors, lasers, and magnetic resonance imaging, make use of qualities derived from the rules of quantum mechanics, or at the very least must take these properties into account while designing their devices. While quantum mechanics has already been incorporated into current information and communication technologies (ICT), the potential of quantum capabilities is significantly larger than that of conventional resources. Over one hundred years have elapsed since the discovery of quantum theory. This year marks the centenary of Einstein's Nobel Prize in physics for his description of the photoelectric effect, a watershed moment in the field's progress. Since Einstein's time, systems capable of managing and utilizing quantum resources have advanced greatly in sophistication. Similar advancements in quantum theory have given non-classical applications in disciplines as varied as communications, information processing, modeling, and sensing a new lease of life. Many developments have also been achieved in experimental platforms, such as quantum photodetectors, superconductors, and trapped ions.

With the maturation of quantum information science (QIS), responses from ICT leaders have ranged from apathy to enthusiasm. Others believe quantum computing is a one-trick pony that would destroy our present security infrastructure by exploiting flaws in public-key encryption, as some have suggested. Others think that quantum information systems (QIS) will give a solution to practically every challenge in information and communications technology, including a road to faster-than-light communications and a mechanism for handling large amounts of data.

The promise of QIS is somewhere in the middle: it offers a broader range of possible applications than some believe, but its capabilities are not endless. We do not yet know what a humongous blunder quantum computer capable of cracking encryption would look like, and QIS cannot defy relativity or address all of our data-related issues.

2.2 History of Quantum Technology

Quantum science and technology is a hot commodity worldwide, with governments, large IT businesses, and investors throughout the globe splashing amount in research and development.

Quantum Technology, from theory to revolution

Observations were obtained in the early 1900s that were not compatible with the old, classical theories of physics. Throughout one study, researchers concluded that hot, black things generate infrared waves at wavelengths that were not envisaged and that atoms could only emit and absorb light at specified frequencies.

In sought to explain these contradictions, the physicist Max Planck proposed in 1900 that light was emitted in tiny, discrete "packages" - that it was quantized – and that it was quantized. This marked the beginning of quantum physics, which may be the science that explains the universe at the atomic scale. Niels Bohr and Albert Einstein were among the most brilliant scientists of their day, and they were in charge of subsequent growth. As they persisted through the analysis, it became clear that they were on the track of a radical shift rather than just a correction to classical physics. Even though many of the repercussions of quantum theory were not fully appreciated until much later, the theory was substantially complete by the 1930s.

The field of quantum physics has had a significant influence on civilization. Researchers were able to develop both the laser and the transistor by using the properties of quantum physics in light and materials, respectively. Such technologies serve as the foundation for informatics as a whole – computers, the internet, and a slew of other innovations – and have played a significant role in shaping contemporary civilization. This was the beginning of the first quantum shift.

However, even though researchers learned how to exploit some quantum properties, it was long believed to be difficult to foresee discrete quantum systems, such as individual atoms, electrons, or light particles, because of the nature of quantum theory (photons). In the 1980s, physicists were successful in creating techniques for monitoring and regulating individual atoms and photons, work that culminated in the awarding of the Nobel Prize in Physics in 2012 for the field. Parallel to this, other researchers produced electrical components made of semiconductors and superconductors, which allowed them to manipulate individual electrons in a controlled environment.

The ability to control individual quantum systems has paved the way for a second quantum revolution, which will open up a slew of wholly new possibilities. Goals today include developing incredibly fast computers, intercept-proof communications, and hyper-sensitive measurement technologies, among other things.

Having spent many years on basic research, we see the applications of quantum technology becoming more accessible. Researchers, policymakers, and corporate executives are all beginning to see that quantum technology has the potential to alter our society dramatically. Quantum technology is now attracting significant investment from investors throughout the globe.

2.3 Phenomena of Quantum Technology

Some of the implications of quantum physics appear inconsistent with common sense. However, quantum theory has shown to be right on all of the issues that can be independently checked. Quantum technology is a scientific principle that seeks to use the characteristics of quantum physics to make new things conceivable. The most significant occurrences are detailed in further detail below.

Superposition

Things have unique attributes in our daily lives — for example, something can only be in a specific location at a given instant – that help us identify them. However, in quantum physics, uncertainty and chance play a significant role. In the same way that electrons spinning around atomic nuclei may be in two places simultaneously, a luminous object (photon) can go down two separate trajectories at the same time. Superpositions are the technical term for such composite circumstances. Once a measurement is taken, the particle is pushed into one of the various options, with the choice being determined by chance. The idea of superposition is broad and may be applied to various characteristics, including energy, electric charge, and velocity.

Using superposition, it is feasible to store and organize massive amounts of information; for additional information, see Section 3.1, Quantum computers.

Entanglement

Entanglement is a superposition phenomenon that occurs between people or more particles and extends between them. Surprisingly, the entangled state of the particles persists even when a significant amount of distance separates them.

For example, light particles, or photons, may be polarized in either the horizontal or vertical directions. The entangled condition is created by placing two photons in a state that is a superposition of two states: one in which both photons are simultaneously polarized and another in which both photons are vertically polarized. Thus, the polarizability of the atoms is unknown, but they have to have the same polarization as the background light. Afterward, we launch one photon toward the moon. The polarization of the second photon is measured on earth, and the result is either horizontal or vertical, depending on the circumstances. And, even though the moon is so far away and has no network connection with the earth, the photon on the moon quickly acquires the same polarization. Einstein was quite doubtful about this and referred to it as "spooky activity at a distance," but investigations have shown that he was accurate.

Entangled states may be utilized to convey communications that are entirely impenetrable to interception.

The Squeezed States

Heisenberg's uncertainty principle is considered one of the fundamental principles of quantum physics. This statement indicates that the accuracy with which the location and velocity of an item may be determined simultaneously has a limit. Other interconnected variables, such as frequency and time, are subject to restrictions.

In most cases, the uncertainty is distributed evenly between the two variables. However, by controlling the quantum system, it is possible to guarantee that the uncertainty only impacts one variable in the system. This is referred to as being in a squeezed condition. This condition makes it feasible to determine the dependent variable with ultrahigh accuracy, which may be exploited to create highly sensitive measuring devices.

2.4 Four Areas of Quantum Technology

Individual quantum systems can be accurately controlled, allowing the utilization of the phenomena mentioned above. Quantum technology is built to control discrete quantum components of high precision. Among the uses for this technology are encrypted communications, very sensitive measuring techniques, and computational capacity much above what is now available from supercomputers.

According to popular opinion, the field of quantum technology may be separated into four domains. These include quantum computing, quantum simulation, quantum communication, and quantum sensing. The final two are quite near to commercialization, and items have already begun to emerge on the market.

Quantum Computing

Modern computers use bits as their smallest data carriers, and bits can only have two possible values: 0 and 1. However, due to superposition, the quantum equivalent, also abbreviated as a quantum bit or qubit, may have the values 0 and 1 simultaneously as a conventional bit. It is feasible for two qubits to have four different values simultaneously — 0, 1, 10, and 11 — with each additional qubit increasing the number of potential states by double. Thus, a quantum computer with 300 qubits would represent more values than there are particles in the whole universe if there were any. And it only needs 50-60 qubits to surpass the computational capability of today's supercomputers, which is a small amount of data. Additionally, a quantum computer may be used to tackle issues that have a huge number of possible alternative solutions, such as optimization challenges in machine learning and artificial intelligence, among other things. It is also well suited for estimating the properties of big molecules — for example, DNA. As an example, consider the development of novel medicinal goods or materials.

Quantum Simulation

Essentially, a quantum simulation model is a quantum computer that has been carefully created to imitate a certain operation. As a result, it can only solve a restricted number of issues. Alternatively, if you wish to tackle additional issues, you must create a new quantum simulator specifically built to address those difficulties.

Although several very modest instances of quantum simulation have previously been shown, quantum computers have not yet exceeded the performance of conventional computers. Despite this, researchers are making rapid progress and ready to expand up to the level necessary to show what is referred to as transcendence, which is the ability to resolve issues that are far beyond the grasp of even the most potent computing environment. It is envisaged that practical applications for quantum simulation would be available within five years.

Quantum Communication

We live in an internet-based civilization, including online banking, digital clinical notes, web-based shopping, and other services that rely on safe data transfer. Modern encryption is based on issues that are assumed to be easier to visualize, such as determining the prime number factors that have created a given extremely big number and are used to protect sensitive information.

On the other hand, cracking today's cryptography will be child's play if the quantum computer makes its debut. On the other hand, quantum technology does provide a solution - it allows for the safe transfer of encryption keys using quantum communication. This is the only known technique that can ensure that an outsider will not decrypt the encrypted communication and read its contents.

The encryption algorithm is a code that the receiver must know to decipher the encrypted communication sent to him. The transmitter uses individual photons to communicate the secret technique to the receiver.

It is impossible to quantify a photon without having it distorted means that you can be certain of detecting whether or not an adversary has tried to grab the secret technique.

Quantum Sensing

The extent to which we can measure things and how accurately we can measure them limits our understanding of the world and our technical progress. A new technique for measuring forces, gravity, electric fields, and other phenomena is being developed by researchers learning to employ molecular orbitals such as photons and electrons as sensors. As a result, measurement capabilities are pushed well beyond previously conceivable. Examples include measuring methods that can measure forces as faint as the gravitational pull between two persons on opposite sides of the Eastern hemisphere.

2.5 Myths about Quantum Physics

For hundreds of years, it seemed that the rules of physics were fully predictable. If you knew precisely where every component was, how fast it was traveling, and what the attractions were both of them at any given moment, you could easily predict what they'd be doing and what they'd be doing at any time in the future. However, this is not possible. There was no built-in, intrinsic ambiguity to the principles that controlled the Universe, from Newton to Maxwell, in any form. All of your limitations resulted from your inadequate information, statistics, and calculation capabilities.

All of that was different a little more than a century ago. From radioactivity to the photoelectric effect to the behavior of light when it is passed through a double slit, we realized that, under many scenarios, we could only recognize the likeliness that various outcomes would occur as a result of the quantum nature of our Universe. This realization led to the development of quantum theory. There have, however, been numerous myths and misunderstandings that have developed in tandem with this new, paradoxical view of reality.

Here are the some myths about quantum physics

- **In quantum physics, uncertainty is the key concept**

The 'uncertainty principle' was developed by Heisenberg. Authors Margaux Khalil and Janet Rafner donated the material for this concept.

Wrong! Probably the most exact scientific subject ever established by humans, quantum physics is the study of subatomic particles. It can forecast some attributes with exceptional precision, down to ten decimal places, which further studies have shown to be accurate.

This myth may be traced back to Werner Heisenberg's "uncertainty principle," partially responsible for its creation. He demonstrated that there is a limit to how precisely two values — for example, the speed of a particle and its location – can be measured simultaneously in the same place. When quantum physics is utilized to compute other variables, such as energy or the magnetic property of atoms, the accuracy with which the results are obtained is remarkable.

- **Quantum physics is impossible to visualize**

Wave functions, superimposed states, probability amplitude, and complex numbers, to name a few of the objects described by quantum physics, are often described as "strange" and difficult to visualize: wave functions, superimposed states, probability amplitude, and complex numbers, to name a few examples. Individuals often assert that they can only be understood via the use of mathematical equations and symbols. Despite this, when we teach and popularize physics, we physicists are always creating representations of it. We utilize graphs, drawings, metaphors, projections, and various other tools to communicate. It's a good thing, too, since students and even seasoned quantum physicists like us need a mental picture of the things being modified to understand what is going on. The authenticity of these pictures is a point of contention since it is impossible to portray a quantum entity on a computer screen adequately. The Physics Reimagined research team is collaborating with designers, illustrators, and video producers to "create" quantum physics in all of its manifestations, including folding activities, graphic novels, sculptures, 3D animations, and so on.

- **Even experts are baffled by quantum physics, which they admit**

One of the field's most prominent figures, Richard Feynman, once said, "I believe I can confidently declare that nobody knows quantum mechanics." Nevertheless, he promptly followed up by saying, "I will explain to you how nature acts." Niels Bohr, one of the discipline's founding fathers, provides a concise summary: "Those who are not surprised when they first encounter quantum theory cannot conceivably have comprehended it."

Scientists who manipulate quantum formalism are aware of what they are doing, explaining their actions to others. All they have to do now is convert their intuitions to this new area and its inherent contradictions to be successful.

- **A small group of bright theorists developed the whole notion of quantum physics**

However, the whole historical record of quantum physics illustrates the polar opposite. From the very beginning, lab tests produced unanticipated outcomes such as the photoelectric effect and black-body radiation, as well as the electroluminescence spectrum of atoms. Whenever Albert Einstein, Max Planck, Niels Bohr, and others attempted to give explanations, great thinkers like Albert Einstein and others joined the picture. Further basic tests proceeded, notably electrons that skipped in an unusual manner off nickel, silver atoms that wandered unusually when exposed to a magnetic field, a flawless conduction metal at low temperatures, and so on. Models and theories began to develop once again, including parallelism, spin, and conductance, among other things. The very fruitful "back and forth" interactions between theory and practice constitute the foundation upon which physics is formed. Except for a few rare instances, experiments are usually carried out first.

- **Einstein was quantum physics' deadliest enemy**

Poor old Albert Einstein is sometimes represented as having been an outspoken opponent of quantum physics, most likely because of his famous phrase, "God does not play dice with the cosmos," which states that "God does not play dice with the universe." Nonetheless, he was not opposed to it, and in fact, he was the one who invented it! Build on the study of Max Planck, and Einstein produced his seminal piece "On a Heuristic Viewpoint Troubling the Emission and Transition of Light" in 1905, which is considered his fundamental work. He postulated that light is composed of photons, which are tiny, discrete, and quantized particles of matter. In reality, this achievement earned him the Nobel Prize, not his contributions to the theory of relativity. That reputation was most likely obtained by Einstein due to his disputes with Niels Bohr, particularly on the concepts of hermeneutics and quantum reality, since he was adamantly opposed to the notion of nonlocality. His claims were later disproved by tests on resonance and violation of Bell's theorem, demonstrating the lack of hidden variables. When it came to quantum physics, Einstein had no doubts about its importance; he has had a few reservations about certain of its consequences, particularly those relating to locality.

- **Quantum physics has no practical application**

Quantum physics is possibly the most valuable field in contemporary physics since it has allowed scientists to influence light, atoms, and electrons after they have learned how they operate. Quantum physicists were responsible for inventing several technologies, including lasers, MRI in hospitals, LEDs, storage devices, hard drives, and, most all, the transistor and electronic components.

2.6 Difference between Quantum and Classical Physics

Classical Physics

Conventional physics is committed to the highest standards of Newtonian mechanics, which are at the heart of the discipline. Traditional physics comprises Newtonian mechanics and thermodynamic parameters, while also wave model of optics and Maxwell's electromagnetism theory. These theories may be used to describe an enormous variety of macroscopic events that occur throughout the spiritual realm. This kind of theory fails catastrophically when applied to events occurring in the atomic and nuclear regimes, such as proton atom sprinkling or the transport of electrons in a semiconductor, for example. When it comes to scientific theories, quantum mechanics is the most successful one that has ever been developed, and it has fundamentally transformed our understanding of the universe. We can see how classical physics has fallen short of its goals using black body radioactivity and the photoelectric effect.

Because they presented explanations for both observations that were based on quantum field theory, Max Planck and Albert Einstein are regarded as the founding fathers of Quantum Physics.

Quantum Physics

For more than many decades, scientists have investigated quantum physics in precision, day in and day out. Werner Heisenberg, Max Born, and Pascual Jordan were the first to describe it as composite mechanics; later, Louis de Broglie and Erwin Schrödinger established it as quantum physics; and finally, Fermi-Dirac and Bose-Einstein established it as quantum survey information of component atoms. It is now known as quantum systems. During the 1930s, Dirac developed his general relativity quantum mechanics, which combined relativistic physics with quantum mechanics to create a new theory. The Uncertainty Principle is considered to be the cornerstone of Quantum Physics. The importance of randomness in tiny physical processes dispels the popular belief that the cosmos is organized. When considered in the classical sense, a quantum world is unexpected, and it completely discredits the notion of an objective cosmos. Scientists continue to rely on the Copenhagen understanding of the quantum mechanical paradigm that is now most commonly recognized by the scientific community. Quantum theories indicate the existence of a cosmic spirit that permeates the universe and the interdependence of people in the global society. Quantum philosophy is comprehensive, and it has the potential to transform our universe completely.

Difference between Classical and Quantum Physics

Mechanics is concerned with macroscopic components, while Quantum Mechanics is concerned with tiny particles.

Classical mechanics is based on Newton's laws of motion, which are still used today. When studying quantum mechanics, it is important to consider Heisenberg's cosmological constant and the de Broglie idea of the dual nature of matter.

Classical mechanics is based on the electromagnetic wave theory developed by Maxwell. According to it, any quantity of energy may be constantly released or absorbed without interruption. It is founded on Planck's quantum theory, which states that only discrete energy values may be emitted or absorbed in the presence of other discrete energy values. According to Classical Mechanics, the state of a system is described by defining all of the forces operating on the particles in that system. It also keeps track of particle counts, locations, and velocities (moment). In this case, the future state may be anticipated with confidence. Quantum Mechanics calculates the likelihood of detecting the particles at different points in space using various parameters. R is a constant.

Chapter 3 Quantum Physics and Experiments

3.1 Rutherford Experiment

The Rutherford Atomic Model, popularly described as the plum pudding model, was proposed by J. J. Thomson but was insufficient in explaining some research data related to the arrangement of atoms of elements. Using the results of a study done by Ernest Rutherford, a British physicist, he postulated the atomic structure of elements. He named it the Rutherford Atomic Model based on his findings. Rutherford proposed a model in which he projected high-energy showers of gamma-particles from a radioactive source against a thin sheet of gold (with a thickness of 100 nanometers). To observe the deflection of the -particles, he set a fluorescent zinc sulfide barrier together around the thin sheet of gold and observed the results. Rutherford made significant findings that directly conflicted with Thomson's atomic hypothesis.

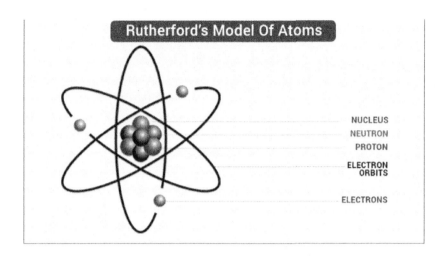

Rutherford Atomic Model

Rutherford hypothesized the atomic structure of substances based on the facts and conclusions presented above. As per the Rutherford atomic hypothesis, the following is true:

- The positively charged particles and the majority of the mass of an atom were condensed into a very tiny volume. The nucleus is the area of an atom that was named for him.

- The Rutherford model stated that the nucleus of an atom is surrounded by electrons that are negatively charged. He also asserted that the electrons around the nucleus travel in a circular circuit everywhere at very high speeds, as indicated by him. Orbits were the names he gave to these circular pathways.

- In addition, a strong electrostatic force of attraction holds electrons, which are negatively charged, and the nucleus, which is composed of a heavily compacted amount of positive charge, around each other.

Rutherford's Alpha Scattering Experiment Findings

Rutherford came to the following conclusion as a result of his assertions:

- A large proportion of the -particles bombarding towards the gold sheet went through it without being deflected. As a result, the majority of the volume in an atom is occupied by unoccupied space.

- Because some of the states attained were deviated by the gold sheet at very tiny angles, the positive charge in a particle is not divided equally across the atomic structure. The positive charge of an atom is localized in a relatively tiny amount, making it difficult to detect.

- The deflection of just a few -particles was approximately 180 degrees; hence, only a few -particles were redirected back in the opposite direction. Consequently, the positively electric charges take up a relatively little portion of the overall volume of an atom when measured to the whole dimension of an atom.

Limitation in Rutherford's Model

The Rutherford Atomic Model has certain limitations. Even though the Rutherford atomic model was based on actual evidence, it could not explain many phenomena.

- Rutherford hypothesized that electrons circulate about the nucleus in definite trajectories termed orbits, which are defined by the laws of physics. Following Maxwell's law, accelerating charged particles create electromagnetic radiation, and therefore an electron rotating around its nuclear nucleus should also release ionizing energy. This emission would convey energies from the mobility of the electron, but it would do so at the expense of the orbits of the electrons, which would shorten as a result.

- Eventually, the electrons would collide with one another in the nucleus. It has been determined via calculations that an electron would collapse in the nucleus in less than 10-8 seconds if the Rutherford model were followed. As a result, the Rutherford model did not agree with Maxwell's theory and could not be used to explain the stability of an atom.

- Including some of the disadvantages of Rutherford, the model was the absence of any description of the configuration of electrons in an atom, which rendered his theory unfinished.

- While the first generation of atomic models was imprecise and attempted to understand some study data, they served as the foundation for subsequent breakthroughs in the field of quantum systems.

3.2 Quantum Tunneling

A quantum mechanical tendency in which particles have a finite probability of bridging binding energy, such as the energy dissipated a binding between another substances. Although the molecule's energy is far less than the imaginary line drawn, it is known as the energy barrier effect. A particle can never traverse an energy barrier with a higher energy level than the particle's energy level in classical mechanics, which means that quantum tunneling has no parallel in classical mechanics. The production of alpha rays shows quantum tunneling during radioactive decay. Even though these particles are firmly bonded to the nucleus and do not possess as much momentum as the bond, they have a limited chance of escaping from the nucleus. This effect is used in constructing transistors and a large number of diodes. Tunneling, which has no analog in conventional physics, is a significant outcome of quantum mechanics and deserves to be discussed further. Take the issue of a molecule with energy E located in the inner area of a one-dimensional probability V (x). In physics, a hypothetical well is defined as a possibility that has a different amount in a specific space area than in the surrounding communities. The particle stays in the well in classical mechanics if E is less than V (the ultimate depth of the barrier); if E is more than V, the particle is expelled from the well. In quantum mechanics, however, the issue is not as straightforward. Particle escape is still possible even if the particle's energy E is lower than the barrier's height V, albeit the likelihood of escape is low until E approaches the barrier's height. In such an instance, the particle may tunnel through the potential barrier and emerge with the same energy E that it had when it went through it. Tunneling is a phenomenon that has a wide range of relevant applications. A form of radionuclides in which a nucleus produces an alpha particle, for example, is described by this term (a helium nucleus). According to the quantum interpretation proposed concurrently by George Gamow and Ronald W. Gurney and Edward Condon is 1928, the alpha particle is restrained before decaying by a potential before decaying into other particles. When dealing with a certain nuclear species, it is feasible to determine the energy E of the released alpha particle and the average lifespan of the nucleus before decaying. The lifespan of the nucleus is a likelihood of tunneling through the barrier; the shorter the longevity, the greater the chance of tunneling through the barrier.

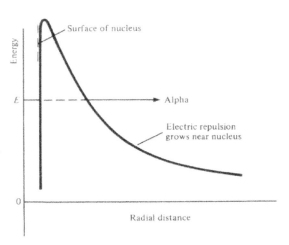

Potential barrier around a uranium nucleus presented to an alpha particle. The central well is due to the average nuclear attraction of all the nucleons and the hill is due to the electric repulsion of the protons. Alpha particles with energy E trapped inside the nuclear well may still escape to become alpha rays, by quantum mechanically tunnelling through the barrier.

Suppose one makes generalizations about the general structure of the potential function. In that case, it is feasible to construct a connection between an E that is appropriate to all alpha emitters without resorting to complex mathematics. This hypothesis, which the experiment has confirmed, demonstrates that the likelihood of tunneling is particularly sensitive to the value of the parameter E.

The value of E ranges between around 2 and 8 mega electron volts, or MeV, for all known alpha-particle emitters (1 MeV = 10 electron volts), depending on the source. This means that the value of E fluctuates only one factor of four times, while the range is from about 10^{11} years down to around 10^{-6} seconds, a factor of 10^{24}.

It would be impossible to explain this sensitivity to the value of E using a theory other than quantum mechanical tunneling for accounting for this sensitivity.

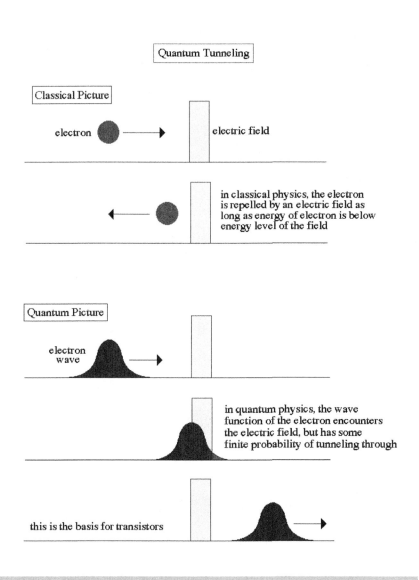

3.3 Schrodinger's Cat

In a classic thought experiment by Austrian scientist Erwin Schrödinger, the cat Schrödinger's cat plays the role of the main character. The item was used to demonstrate a hypothetical contradiction that might develop when the Copenhagen interpretation of quantum physics is applied to common things, which he explained in detail.

Niels Bohr and Werner Heisenberg began working with the Copenhagen interpretation between 1925 and 1927, and it is still in use today. It asserts that physical systems controlled by quantum mechanics do not have definitive qualities until they are measured, contrary to conventional wisdom. A superposition of various states, on the other hand, means that an item may exist in several states at the same time, each of which has a separate set of qualities. When a quantum item is in two locations simultaneously, it is said to be traveling at two distinct speeds simultaneously. These states have varying probabilities, which are defined by the wave function of the item. In doing a measurement, the numerous alternative states are focused on a single number, which is referred to as "collapse."

Schrödinger's assumed research project attempts to link the qualities of a tiny quantum-mechanical system (a radioactive substance) to the attributes of a commonplace, huge item (a radioactive substance) (a cat).

The experiment is set up in the following manner (as seen on the right): A cat has been imprisoned in a room containing radioactive material. A Geiger counter is a device that determines whether or not an atom of radioactive material decays and releases radiation. If radiation is detected in the chamber, a flask containing lethal hydrogen cyanide is crushed, resulting in the feline's death. The cat can survive even if no atom decays and no radiation is detected. It is impossible to tell if the cat is alive or dead until one looks inside the box, that is, until one measures the system's condition. The cat's wave function after a certain period will depict it as being equal parts alive and dead if the chance of radioactive decay during that period is, for example, 50 percent. According to Schrödinger, this was inconceivable, and he decided that the Copenhagen interpretation had to be incorrect.

Other interpretations of quantum mechanics have been offered, such as the many-worlds interpretation, according to which reality is divided into multiple branches for each potential state of existence. Even now, philosophers are debating the philosophical challenge raised by the contradiction.

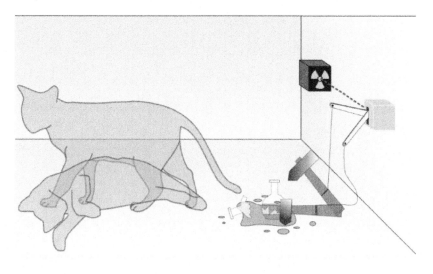

3.4 Quantum Field Theory

In its simplest form, quantum field theory is the exploration of the quantum characteristics of fields, where "fields" refer to any physical phenomena that extend over space and time. The acoustic radiation and the oscillation of crystals are examples of extended physical phenomena that are commonplace in our universe. Consequently, quantum field theory is fundamental to many areas of theoretical physics, such as high-energy particle physics and condensed matter physics, for which it provides a foundation. Indeed, research into quantum mechanics began shortly after the theory was conceived only a hundred years ago.

In one sense, quantum field theory has advanced to a mature state and has been applied to a wide range of phenomena; additionally, the results of computations in quantum field theory have been shown to duplicate experimental data reliably. For illustrate, the Standard Model of particle physics itself, which explains the interactions of basic particles below the energy scale that is presently accessible to humans, is one kind of quantum field theory among many other types of theories. Several mathematics areas were partially established due to different inspirations derived from the research into quantum field theory, namely the Seiberg-Witten theory of manifolds in four dimensions, which is an example of such a subject.

Theoretical physics' quantum field theory is still a very new, immature, and evolving topic of study. One evidence of quantum field theory's relative freshness is that there hasn't been any consensus on a mathematical language that can explain what physicists do with it, even though mathematicians have attempted to articulate different portions of the theory in the past. This is in marked contrast to the relatively with, for example, quantum physics and general relativity, for which we have accurate mathematical theories to explain the phenomena. Scientists understand a set of principles that enables them to calculate experimentally observed values. However, the framework that encompasses these rules is still a long way off in development.

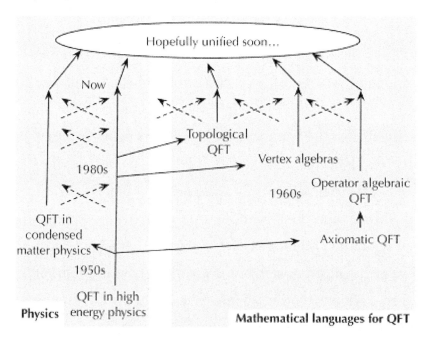

3.5 Quantum Gravity

When it comes to gravitational attraction, Quantum Gravity refers to any theory that represents gravity in situations where quantum phenomena cannot be ignored. At the current time, there should be no such theory that is widely accepted and supported by empirical evidence in the field. Thus, the phrase "Quantum Gravity" refers to an issue that has yet to be solved rather than a particular theory. Diverse studies lines, each at the forefront of this change, are being pursued to find potential answers to the issue. These preliminary quantum-gravity explanations are either incompatible research lines or additions to a shared objective of discovering a physically accurate theory, depending on who you ask. There are several implications of the search for a good quantum theory of gravity on basic difficulties. It is commonly characterized as the most significant outstanding topic in fundamental physics or the "Holy Grail" of modern theoretical physics since it touches many of these issues.

Problem in Quantum Gravity

It is general relativity, which Albert Einstein created at the beginning of the twentieth century. It has profoundly altered our understanding of space and time, that is, the modern theory of gravity. Additionally, quantum physics has profoundly altered our understanding of matter, energy, and cause-and-effect relationships. These two hypotheses have triggered a profound fundamental revolution in physics, which is now extremely strongly supported by empirical evidence. However, they seem irreconcilable, at least at first glance, since each of the two theories is developed based on premises that are openly denied by the other explanation. Due to the enormous advancement of scientific knowledge that the physics of the twentieth century has brought us, we are still in a state of perplexity about what the fundamental conceptual elements are for comprehending the physical universe. For want of a better phrase, the scientific revolution sparked by generic relativity and quantum mechanics at the decade of the twenty century has not yet been completed, and a new synthesis is needed. This new synthesis should be based on quantum gravity, which combines general relativity with quantum mechanics.

Applications in the physical world

The following are some examples of phenomena in which quantum gravity significantly influences.

- **The Microscopic Structure of Space time**

In the first place, we should be able to see quantum gravity effects if we can characterize the geometry of temporal and spatial at the Planck scale, which we cannot currently do. According to many considerations, the Planck length may be seen as a limit to the infinite divisibility of space or as a minimum length. Any effort to measure smaller distances would, according to intuition, result in the buildup of this much radiation in too tiny a space, with the outcome being the formation of an artificial black hole, thus removing the area from view altogether. It might be possible to complete the trio of basic scales in Nature by determining a minimum length. This would be in addition to the speed of light, the maximum velocity of an object, and the Planck constant, the smallest amount of action transferred between two systems.

- **Prehistoric cosmology**

According to the widely accepted cosmological model, the universe was very dense and hot in the prehistoric period. The Big Bang is a solitary point of infinity in terms of density, temperature, and curvature that is reached by extrapolating the model backward from the present day. On the other hand, this ultimate extrapolation is very probably inaccurate because quantum gravitational interactions become dominating when the cosmos is pretty dense and hot, and these consequences are not included in the conventional model. It is necessary to develop a quantum theory of gravity to account for these phenomena and investigate the first moments in the history of our universe. It is possible that the unique Big-Bang point was never achieved. The universe's present expansion preceded a collapsing phase and a "Big Bounce," according to certain modern theories of gravity (particularly loop quantum gravity, which is discussed below). A major hope for discovering traces of quantum gravitational phenomena is in this cosmological context, as markings of early physical world manifestations left in the cosmic background radiation, which is currently under intimate assessment, or in the reference gravitational wave irradiance, which is intended to be observed within the next decade. One of the major challenges of discovering visible signs of quantum gravitational phenomena is the difficulty separating them from other types of gravitational phenomena.

- **Black holes**

Quantum gravity is expected to have a role in various areas of black-hole physics, including the formation and evolution of black holes. First and foremost, it should provide a comprehensive knowledge of the radiant heat that black holes are projected to emit, as estimated by Stephen Hawking in the early 1990s. Second, Hawking's study demonstrates that black holes contain immense entropy, with a value of 10^{77} for a black hole with a mass equal to that of the sun. Even by gigantic thermodynamic standards, the statistical mechanical explanation for this large quantity is not well understood. Third, quantum gravity is anticipated to replace the endless singularity predicted by general relativity at the heart of black holes with a more physically plausible image consistent with current knowledge. The theory should also explain what occurs after a black hole's Hawking evaporation, which is the last stage of the process.

- **Astrophysical effects**

Numerous cosmological quantum-gravitational consequences have been proposed, including one involving a black hole. Although none have been detected so far, several computations predict that they may be observed in the not-too-distant future. Examples include a slight influence of the speed of light on the color of the light, which is induced by Planck-scale granularity of space and maybe seen experimentally. Because of the bigness of the Planck scale, the effect is very tiny; nevertheless, if the influence is cumulated over a very long extraterrestrial route taken by the light, it may become observable over time. Observations are being conducted to see whether or not this prediction is correct.

3.6 Quantum Entanglement

Newtonian physics is effective up to a point, after which a different set of principles, such as Einstein's relativity and even more fundamental Quantum Mechanics, take over. Consider the following analogy: consider a mound of sand, the grains of sand that make up that pile, and the silicon atoms that make up the sand; they are both the same thing and distinct from one another.

Quantum Entanglement is a state in which two systems (a platform is most often an electron or a photon) are so closely correlated that gather data regarding a particular platform's "state" (for example, the position of the electron's spin, say "Up") will give immediate information about the other system's "state" (for example, the direction of the electron's spin, say "Down," 100 percent of the time), no matter who the other two components are located. "Instant" and "regardless of how far apart they are" are significant. Because the state of a solution can be controlled until it is evaluated, and because the transfer of information disregards the customary physics rule, which states that evidence cannot be relocated close to the speed of light, investigators such as Einstein have been perplexed by this phenomenon for decades. However, since the 1980s, researchers and experimenters have been able to demonstrate that entanglement can be established using both photons and electrons.

In certain cases, two subatomic particles (the electrons) may be prepared so that a single wave function can describe their wave functions. In one example, the decay of an entangled father particle with zero spin results in two entangled daughter particles with equal but opposing spins, which may be used to achieve entanglement.

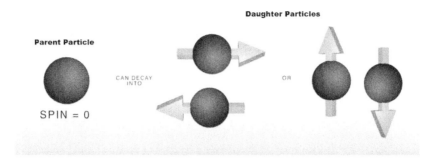

No matter how far away the two daughter particles are from one another, their wave functions (when measured) will remain equal and opposing if they do not interact with anything. Scientists have discovered that the information is not determined from the moment of entanglement, as was previously believed. Instead, only when one has measured the information contained in one particle is communicated to the other particle, which occurs at a rate faster than the speed of light itself. So the information does flow at this speed, but we have no way of controlling it; this inability to control the information limits the possible applications of Quantum Entanglement, such as delivering a message or other information faster than the speed of light.

Usage of Quantum Entanglement

Many applications will be able to make use of this unique physical trait, which will have a substantial influence on our present and future. Entanglement can allow quantum cryptography, super dense coding, communication that is perhaps quicker than light speed, and even teleportation.

Quantum computers have the ability to tackle time- and processing-intensive challenges in a variety of sectors, including finance and banking, as a result of their potential. As a result of quantum entanglement, such computers might save time and computing power by reducing the time and processing power required to process information transmission between their qubits.

Chapter 4: Black Holes

4.1 Black Hole

A black hole is a region in space where the gravitational attraction is so powerful that nothing, not even light, can escape its interior.

Although black holes were initially imagined in the creative minds of theoretical physicists, they have since been discovered in the hundreds and millions across the universe. Despite their invisibility, these things interact with and impact their environment in a manner that can be detected. The form of such interaction is determined by the distance between the black hole and the observer: too near, and there is no escape, but farther out, dramatic and spectacular occurrences will occur.

The word "black hole" was first used in print in a 1964 article by Ann Ewing, reporting on a symposium held in Texas in 1963, although she never said who created the phrase. Though fellow Americans Robert Oppenheimer and Hartland Snyder conceived the notion of a collapsed star in 1939, American physicist John Wheeler required a shorthand for 'gravitationally totally collapsed star' and popularized the phrase in 1967. In reality, in 1915, German scientist Karl Schwarzschild solved several of Einstein's fundamental equations (known as the field equations) for an isolated non-rotating mass in space, laying the mathematical foundations for the present image of black holes. Sir Arthur Eddington had worked out some of the essential mathematics in the context of studying work by the Indian physicist Subrahmanyan Chandrasekhar on what happens to stars when they die two decades later, a little before Oppenheimer and Snyder's work. The physical consequences of Eddington's calculations, notably the collapse of enormous stars after they had used up all their fuel to become black holes, were deemed 'absurd' by Eddington himself in 1935 to the Royal Astronomical Society. Despite their seeming ridiculousness, black holes are very much a component of physical reality in our galaxy and across the universe. David Finkelstein made further advancements in the United States in 1958, establishing the existence of a one-way surface encircling a black hole, which has enormous implications for what we will discuss in the next chapters. This surface prevents light from breaking out from the black hole's enormous gravitational pull, which is why it is black. To begin comprehending how this behavior may occur, we must first comprehend a fundamental aspect of the physical world: every particle or item has a maximum moving speed.

The mass of the falling star determines the size of the Schwarzschild radius. The radius of a black hole with a mass ten times that of the Sun would be 30 kilometers (18.6 miles).

After their lifetimes, only the most massive stars—those with more than three solar masses—become black holes. Stars with less mass develop into white dwarfs or neutron stars with less compacted bodies.

Because of their small dimensions and the fact that they produce no light, black holes are difficult to detect firsthand. However, the impact of their massive gravitational fields on neighboring stuff may be "seen." Suppose a black hole is part of a binary star system, for example. In that case, stuff pouring towards it from its partner gets extremely heated and emits abundant X-rays before hitting the black hole's event horizon and vanishing forever. A black hole is one of the stars in the binary X-ray system Cygnus X-1. This binary, discovered in 1971 in the constellation Cygnus, is made up of a blue supergiant and an unseen companion with a mass of 14.8 times that of the Sun Sun, which rotates around each other every 5.6 days.

No stellar origins are thought to exist in certain black holes. According to multiple astronomers, large volumes of interstellar gas are predicted to collect and collapse into supermassive black holes at the nuclei of quasi and galaxies. The energy released by a mass of gas falling rapidly into a black hole is thought to be more than 100 times that generated by nuclear fusion on the same mass. Consequently, the enormous energy output of quasars and certain galaxy systems might be explained by the gravitational collapse of millions or billions of solar masses of interstellar gas into a massive black hole.

Stephen Hawking, a British astronomer, hypothesized the possibility of another kind of no stellar black hole. According to Hawking's hypothesis, during the big bang, a state of extraordinarily high temperatures and density in which the universe began 13.8 billion years ago, multiple small primordial black holes, maybe with a mass equal to or less than that of an asteroid, might have been generated. Like the larger substantial ones, these so-called small black holes lose mass over time and eventually vanish due to Hawking radiation. The Large Hadron Collider might create many small black holes if certain theories of the cosmos that need additional dimensions are right.

4.2 Hawking Radiation

Hawking radiation is a kind of gamma-ray that radiation supposedly released from just beyond a black hole's event horizon. In 1974, Stephen W. Hawking postulated that subatomic particle pairs (photons, neutrinos, and certain heavy particles) that naturally occur near the event horizon might result in one particle fleeing the black hole's vicinity while the other, of negative energy, sink into it. The flow of negative-energy particles into the black hole diminishes its mass until it vanishes in a last explosion of Radiation.

Hawking radiation is a term used to describe hypothetical particles created by the border of a black hole. As a result of this

Radiation, black holes' temperatures are inversely proportional to their mass.

To put it another way, a black hole should shine brighter the smaller it is. Hawking radiation is a prediction supported by integrated general relativity and quantum mechanics models, even though it has never been directly seen. It is named after prominent scientist Stephen Hawking, who argued for the possibility of black hole explosions in a paper titled Black hole explosions? Released in 1974.

If shown to be true, Hawking radiation would suggest that black holes may release energy and hence reduce in size, with the smallest of these immensely dense objects bursting swiftly in a puff of heat (and the largest slowly evaporating over trillions of years in a cold breeze).

What causes Hawking Radiation to be produced by black holes?

The actual mechanism of particle emission from near a black hole's event horizon is highly complicated, requiring a thorough grasp of quantum field theory mathematics.

It's typically defined as the effect of gravity separating twin 'virtual' particles that originate spontaneously from the vacuum. Normally, they would recombine and cancel out, but the divide in this situation allows one-half of each pair to escape as real Radiation.

In actuality, Hawking's popular description of the mathematics portrays a pair of transitory virtual particles impacted by severe gravity, with one-half of the pair eliminating mass from the black hole owing to the particle's negative energy provided by extreme gravity.

4.3 What Is the Size of Black Holes?

Black holes may be very large or very tiny. The tiniest black holes, according to scientists, are as tiny as one atom. These black holes are really small, yet they have the bulk of a mountain. The quantity of substance, or "stuff," is measured in mass.

"Stellar" black holes are different kinds of black holes. It has a mass up to 20 times that of the Sun. There might be a large number of star mass black holes in Earth's galaxy. The Milky Way is the name of our galaxy.

"Supermassive" black holes are the biggest. These black holes have a combined mass of more than a million suns. Scientists have discovered confirmation that at the heart of every giant galaxy lies a supermassive black hole. Sagittarius A is the name of the supermassive black hole in the heart of the Milky Way galaxy. It has a mass of around 4 million suns and would fit within a massive ball big enough to house a few million Earths.

4.4 Stellar black holes

When a star reaches the end of its fuel supply, it may collapse or fall into itself. The new core of smaller stars (up to three times the mass of the Sun Sun) will become a neutron star or a white dwarf. However, when a bigger star collapses, the star continues to compress, becoming a stellar black hole. Individual stars collapse into black holes, which are relatively tiny yet very dense. One of these things has more than three times the mass of the Sun Sun packed inside a city's circumference. It causes a massive amount of gravitational force to be exerted on items in the vicinity of the object. The dust and gas from their surrounding galaxies are subsequently consumed by stellar black holes, causing them to expand in size. The Milky Way "holds up to 100 million black holes," according to a study by UV Irvine academics.

4.5 Intermediate Black Holes

Scientists believed that black holes only came in two sizes: tiny and huge, but a new study suggests that midsize, or intermediate, black holes (IMBHs) may occur. When stars in a cluster collide in a chain reaction, such things might arise. Several of these IMBHs growing in the same area might ultimately collide in the galaxy's core, generating a supermassive black hole. In the arm of a spiral galaxy, scientists discovered what seemed to be an intermediate-mass black hole in 2014. In 2021, scientists used an old gamma-ray burst to discover one.

These IMBHs may occur in the center of dwarf galaxies, according to further study from 2018. Observations of ten such galaxies (five of which had previously been unknown to science until this recent study) discovered X-ray activity, typical in black holes, indicating the existence of black holes with masses ranging from 36,000 to 316,000 solar masses. The data comes from the Sloan Digital Sky Survey, which looks at over a million galaxies and can identify the kind of light commonly seen emanating from black holes scooping up surrounding material.

4.6 The creation of Giants

- **Supermassive Black Holes**

Small black holes abound in the cosmos, while supermassive black holes, their relatives, rule. These gigantic black holes are millions or billions of times as big as the Sun Sun, yet have around the same size. Black holes are assumed to exist at the heart of almost every galaxy, including our own Milky Way. Scientists are baffled as to how such massive black holes form. Once born, these giants collect mass from the dust and gas surrounding them, which is abundant in the core of galaxies, enabling them to grow even larger.

Hundreds of thousands of small black holes may join together to become supermassive black holes. Large gas clouds, crushing together and quickly accumulating mass, might also be blamed. A third possibility is the collapse of a stellar cluster, a collection of stars that all collide simultaneously. Fourth, enormous groupings of dark matter might produce supermassive black holes. We can see dark matter because of its gravitational impact on other things, but we don't know what it's made of since it doesn't radiate light and can't be detected directly.

Sagittarius A*, a supermassive black hole at the center of the Milky Way Galaxy, is one such example. Observations of stars around the location of Sagittarius A* reveal the existence of a black hole with a mass of almost 4,000,000 Suns. (These findings earned American astronomer Andrea Ghez and German astronomer Reinhard Genzel the Nobel Prize in Physics in 2020.) Other galaxies have also been found to feature supermassive black holes. The supermassive black hole at the center of the M87 galaxy was imaged by the Event Horizon Telescope in 2017. That black hole has a mass of 6.5B Suns yet is just 38 billion kilometers (24 billion miles) wide. It was the first time a black hole had been directly photographed. The intense effects on plasma spinning at very high velocities at the center of NGC 3842 and NGC 4889, galaxies neighboring the Milky Way, may infer the presence of much bigger black holes, each with a mass comparable to 10 billion Suns.

4.7 What Causes Black Holes to Form?

The tiniest black holes, according to scientists, formed when the universe started. Stellar black holes are created when the core of a massive star collapses in on itself. When this happens, a supernova occurs. A supernova is a star that explodes and sends a portion of its mass into space. Supermassive black holes, according to scientists, were created at the same time as the galaxy in which they reside.

4.8 How Do Scientists Know There Are Black Holes If They Are "Black"?

Because high gravity sucks all light into the black hole's center, a black hole cannot be seen. However, astronomers can see how the black hole's enormous gravity affects the stars and gas in its vicinity. Scientists can look at stars to see whether they are circling or traveling around a black hole. High-energy light is produced when a black hole and a star are near. Human eyes are unable to perceive this kind of light. To observe the high-energy light, scientists employ satellites and telescopes in orbit.

4.9 What do Black Holes Look like?

The outer and inner event horizons and the singularity are the three "layers" of a black hole.

The event horizon of a black hole is the border that surrounds the black hole's mouth and beyond which no light can escape. A particle cannot escape the event horizon once it has crossed it. Across the event horizon, gravity remains constant. The singularity, or one point in space-time where the black hole's mass is concentrated, is the inner area of a black hole, where the object's mass is concentrated.

Scientists cannot observe black holes in the same way that they can see stars and other celestial objects. Instead, astronomers must depend on detecting the Radiation that black holes generate when dust and gas are pulled into the dense entities. However, supermassive black holes at the core of galaxies may be obscured by the dense dust and gas surrounding them, obscuring the telltale emissions.

When the matter is dragged into a black hole, it may bounce off the event horizon and be thrown outward instead of sucked into the mouth. The material is accelerated to near-relativistic speeds, resulting in bright jets of material. Although the black hole is invisible, these strong jets may be observed from considerable distances.

The Event Horizon Telescope (EHT) team published the first photograph of a black hole in 2019. While looking at the event horizon, or the region beyond which nothing can escape from a black hole, the EHT discovered the black hole in the heart of galaxy M87. The graphic depicts the loss of photons in a sudden manner (particles of light). It also opens up a whole new field of study into black holes now that astronomers better understand what they look like.

4.10 Binary Black Holes are Being Illuminated

Using the Laser Interferometer Gravitational-Wave Observatory (LIGO), researchers discovered gravitational waves from merging star black holes in 2015.

In a statement, David Shoemaker, a spokesman for the LIGO Scientific Collaboration (LSC), stated, "We now have more proof of the existence of stellar-mass black holes bigger than 20 solar masses – these are things we didn't know existed until LIGO identified them." The measurements made by LIGO also reveal the direction in which a black hole rotates. Two black holes may spin in the same direction or in the opposite way as they spiral around one another.

Binary black holes are thought to arise in two ways. The first hypothesis is that the two black holes formed as a binary simultaneously from two stars born together and died explosively at around the same moment. The partner stars would have had the same spin orientation as each other, and the two black holes left behind would have had the same spin orientation as well.

Black holes in a star cluster sink to the cluster's core and couple up in the second model. In comparison to one another, these partners would have unpredictable spin orientations. The discovery of partner black holes with various spin orientations by LIGO adds to the support for this idea.

4.11 Weird Facts about Black Holes

The only things in the universe that can imprison light by pure gravitational force are black holes. According to scientists, they are thought to arise when the corpse of a big star collapses in on itself, becoming so dense that it warps the fabric of space and time.

And everything that passes over their event horizons, commonly known as the point of no return, spirals aimlessly towards an unknown end. These gigantic cosmic events remain a mystery after decades of inquiry.

Scientists who study them are still blown away by them. The following are ten reasons why:

- **Sucking isn't something black holes do**

Some people believe that black holes are cosmic vacuums that suffocate the space around them, but they are much like any other object in space, although with an extremely powerful gravitational field.

Earth would not be pulled in if a black hole of similar mass replaced the Sun Sun; it would continue to circle the black hole as it does now.

Although black holes seem to be pulling in stuff from all directions, this is a widespread mistake. Companion stars give up some of their mass in the form of stellar wind, which eventually falls into the grasp of a greedy neighbor, a black hole.

- **Einstein did not discover black holes**

Although Einstein's theory of relativity predicts the emergence of black holes, he did not detect them. Instead, Karl Schwarzschild was the first to demonstrate the existence of black holes using Einstein's innovative calculations.

In the same year that Einstein published his theory, the theory of general relativity, he did this. The Schwarzschild radius, a measurement of how thin you'd have to compress any item to form a black hole, is named after Schwarzschild's work.

Long before this, British polymath John Michell predicted the creation of 'dark stars' with gravitational forces so strong that even light could not escape; black holes were not given their worldwide name until 1967.

- **You and Everything Else will be Spaghettified by Black Holes.**

Black holes have the extraordinary power to stretch you into a lengthy strand of spaghetti. This tendency is appropriately referred to as spaghettification.'

It works because of how gravity acts over time and distance. Your feet are nearer to the center of the Earth right now and so more strongly drawn than your head. That difference in the attraction will work against you at severe gravity, such as near a black hole. As gravity's pull stretches your feet, they will grow more and more attracted as you get closer to the black hole's center. They move quickly as they come near. However, since the upper half of your body is further away from the center, it is not moving as quickly. Spaghettification is the end outcome!

- **Black holes have the potential to create new universes**

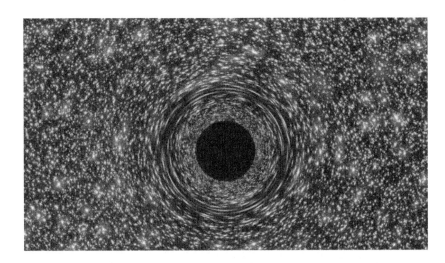

Although it may seem irrational that black holes may produce other worlds – particularly because we have no proof that other universes exist - the underlying idea is still being researched today.

A simplistic explanation of how this works is that today's universe has certain highly favorable circumstances that came together to generate life when you look at the calculations. We wouldn't be here if you changed these criteria even slightly.

The singularity at the center of black holes defies our usual rules of physics, and it has the potential to modify these circumstances and generate a new, slightly different world in principle.

- **The area surrounding black holes is physically pulled in**

Consider space as a crisscrossing grid of stretched rubber sheets. An item dips a bit when it is placed on the sheet.

The larger the item, the deeper it sinks into the sheet. The grid lines become bent instead of straight due to the sinking effect.

Space twists and curls more the deeper the hole you dig in it. Black holes create the deepest wells of all.

- **The ultimate energy factories are black holes**

Black holes are more efficient in generating energy than our Sun. This mechanism is based on the material disc that orbits around a black hole.

The material closest to the event horizon's fringe on the disk's inner border will circle considerably faster than the stuff at the disk's extremely outer edge. It is because the gravitational attraction is larger towards the event horizon.

Because the material is circling and moving so quickly, it warms up to billions of degrees Fahrenheit, allowing mass to be transformed into energy in the form of blackbody radiation.

On the other hand, nuclear fusion turns around 0.7 percent of mass into energy. A black hole's environment transforms 10% of its mass into energy. That is a significant change!

Scientists have even suggested that this kind of energy may one day be utilized to power future black hole starships.

- **Supermassive black hole**

According to scientists, nearly every galaxy, including our own, is thought to have a super massive black hole at its center. These black holes serve as anchors, binding galaxies together in space.

Sagittarius A, the Milky Way's central black hole, is more than four million times more massive than our Sun. Although the black hole, which is over 30,000 light-years distant, is now inactive, scientists think it exploded in an explosion 2 million years ago that may have been visible from Earth.

- **Black holes slow time**

Consider the twin experiment, which is often used to show how time and space interact in Einstein's theory of general relativity:

One twin remains on Earth, while the other travels across space at the speed of light before turning around and returning home. Because the quicker you go, the slower time passes for you, the twin who traveled through space is much younger.

Time will slow down as you approach the event horizon since you are traveling at such high speeds due to the black hole's intense gravitational attraction.

- **Over time, black holes evaporate**

Stephen Hawking anticipated this stunning finding back in 1974. Hawking radiation is named after the great scientist Stephen Hawking.

Hawking radiation disperses a black hole's mass throughout space and time until nothing is left, thereby dying the black hole. Hawking radiation is sometimes known as black hole evaporation because of this.

- **In theory, everything may turn into a black hole**

The sole difference between a black hole and our Sun is that its core is formed of highly dense matter, giving it a powerful gravitational pull. Because of this gravitational field's ability to imprison everything, including light, humans are unable to view black holes. Anything might hypothetically become a black hole.

If our Sun were reduced to a size of just 3.7 miles (6 kilometers) wide, for example, all of its material would be compacted into an extraordinarily tiny region, making it very dense and becoming a black hole. The same concept might be applied to the Earth or your own body.

However, we only know of one mechanism to create a black hole in reality: the gravitational collapse of a big star 20 to 30 times more massive than our Sun.

Chapter 5 Quantum Physics and Daily Life

Quantum physics has been a focus of research and development for researchers and technology titans for more than two decades. With its vast potential, quantum technology can bring the vision of time travel, extraordinary computing, and other possibilities to reality. Although experts have investigated quantum technology for several fascinating studies, we have compiled a list of eight instances in which this innovation is often used in our real activities that are generally overlooked.

- **Electronic Appliances**

Toasters, for example, are examples of how we have been incorporating quantum concepts into our daily lives via electrical appliances. When we are toasting bread in the machine, we see a red glow, which is caused by quantum physics and may be explained as follows: This is because the heating element produces a red color and that heated things generate a glow, which is how quantum physics came to be invented.

- **Fluorescent Light**

Fluorescent bulbs have electrodes that raise the temperature and emit electrons when the bulb is turned on. A little quantity of mercury, which is also contained inside the bulb, gets bombarded due to these electrons. The collisions lead the electrons in mercury to move to a higher quantum energetic state due to the collisions. High-energy electrons tend to return to their lower degree, and whenever they do, photons are released, resulting in the visible light we see today.

- **Quantum Clock**

According to Einstein's theory of relativity, time does not exist in its absolute sense. In other words, time does not advance at the same pace for everyone. For illustration, time seems to move more slowly to a person inside a spacecraft traveling at near the speed of light, yet it appears to move much more quickly to spectators on the ground on Earth. The theory of relativity states that time is relative for each observer and that all various time measurements are equally valid under these conditions.

Time dilation impacts our Global positioning system (GPS and navigational) on a more localized level. When compared to other methods of measuring time, atomic clocks employ quantum physics principles to measure time more precisely than any other method we have previously discovered. It was developed by the National Institute of Standards and Technology in collaboration with the University of Colorado Boulder.

The strontium clock is the most accurate atomic clock ever made. This device is very exact, and in 15 billion years, it will not lose or gain a single second! In addition, the cosmos has been around for 13.824 billion years as of 2018. Mind-boggling! Such atomic clocks might also be connected to a worldwide network, which would allow them to be utilized to measure time regardless of where they were located.

- **Supercharged Computers**

Since its start in the 18th century, computers have gone a long way in functionality (The Analytical Engine). Due to technological advancements, computer processors are now capable of doing millions of computations per second, forecasting the weather, and outperforming humans at chess, among other things.

Basic to computing is the ability of a standard computer to encode data in the form of an arbitrary sequence of binary numbers consisting only of 0s and 1s. The use of "Qubits," bits that are equivalent to both 0s and 1s simultaneously, allows quantum computers to take this to the next level. This implies that it can do specific molecular modeling and number factoring tasks significantly more quickly than a conventional computer.

- **Global Positioning System**

The Global Positioning System, sometimes known as GPS, is a satellite network that has made locating locations and directions easier. Quantum physics is required for the implementation of GPS technology. Time is measured by an ensemble of atomic clocks on each satellite in the GPS constellation, and these atomic clocks use the concepts of quantum theory.

- **Quantum Codebreaking**

Since the beginning of the conflict, attempts have facilitated communication between friends over enemy territory. A particular key was used to decrypt encrypted communications in the most conventional type of cryptography, which is still used today. The most well-known was the Enigma machine, which Germany employed during World War II. The codes generated by the machines were considered unbreakable if they were applied under established processes and standards. Even yet, the Allies could rectify the system and decipher crucial strategic codes throughout the conflict.

Quantum physics provides a possibly indestructible alternative that does not outweigh the previous approaches in place. Quantum Key Distribution (QKD) is a mechanism in which information regarding the key is polarized randomly before being sent. As a result, the photon only vibrates in one direction, such as up-down or left-right, when this occurs. The receiver utilizes the polarized codes to decrypt the key and then uses an algorithm to encrypt the message once they have deciphered the key. It is possible to send encrypted communications across any communication channel, but nobody can decrypt the information even when they have the quantum secret in their possession. Another layer of protection is provided by quantum principles, which mandate that reading the code will eventually modify the photon states, informing the users that a security violation has occurred.

- **Semiconductors**

As the name indicates, semiconductors are elements that have an electrical conductivity that is as there is between conductive elements such as copper and that of a dielectric, such as glass. They are used in electronic devices. A broad variety of currents and voltages may be accommodated by semiconductor devices, making them more suitable in a wide range of technical equipment, including computers, Fluorescent lighting, televisions, mobile phones, and smart gadgets. In reality, our knowledge of quantum physics has played a crucial role in developing all modern devices.

Because electrons are in a quantitative state of superposition, electrical conductivity may be defined as the capacity of electrons to be diffused or deprotonated throughout atoms in a material. Conducting metals are characterized by a high degree of electron delocalization, which allows them to conduct electricity efficiently. Insulators are devices that block electricity from flowing. Semiconductors are crystalline metals with characteristics that lie then between conductors and absorbers and are used in electronic devices. Furthermore, pure semiconductors may be intentionally changed by introducing impurities (a process known as doping) to improve their conducting and shielding characteristics, making them more useful in various applications.

- **Lasers**

Attributed to the reason that the light waves all travel in the same orientation and are coherent, which means that they have similar wavelengths and waveforms, lasers can emit powerful beams of light. When laser light is generated, it is done using a technique known as spontaneous emission. A photon is used to encourage an already enthusiastic atomic electron to drop down to a reasonable quantum energy state, causing the release of two same nature photons that travel coherently. A reflecting chamber is used to repeat this process repeatedly until a large number of photons are coherent and transmitted at the same time. The term laser is an abbreviation for light or other electromagnetic radioactivity, which the technology is all about.

- **MRI**

It is determined by the constitution of a person's vital organs how much hydrogen-dense moisture and adipose molecules are stored in various body parts at different times. Magnetic resonance imaging (MRI) takes advantage of these variations to produce very detailed images.

It is connected with a hydrogen atom's positive electrode proton with a quantum spin, and a revolving charged particle produces a magnetic field. A high external magnetic field causes the magnetic fields of trillions of hydrogen atoms in the body to align, which is not normally the case. When a strong electromagnet is introduced, the axes of the atomic magnetic fields comply. Then, using specially tailored radio frequency pulses, certain hydrogen atoms are briefly knocked out of alignment, allowing the process to continue. In the intervals between pulses, the atoms realign themselves with the external field. During this procedure, portions of the body with a higher concentration of hydrogen atoms may be recognized and distinguished from sections with a lower concentration of hydrogen atoms.

Conclusion

Physics is a natural discipline that studies matter and how it moves through space and time and related concepts like energy and force. In a more comprehensive sense, it is the study of nature to comprehend how the universe works.

Physics employs the scientific method to find the fundamental rules that govern light and matter, as well as the consequences of those laws. It is presumptively true that rules regulate how the world operates and that these laws are at least partially intelligible to humans. It's also commonly assumed that if complete knowledge of the present state of all light and matter were available, those principles might be used to predict the universe's future.

Anything with mass and volume is often referred to as matter. Many of the ideas and rules that describe matter and motion are fundamental to the study of classical physics. For example, the law of conservation of mass stipulates that no mass may be generated or destroyed. As a result, while forming theories to explain natural events, further experiments and computations in physics consider this rule.

Physics seeks to explain how everything around us works, from the movement of small charged particles to the movement of people, automobiles, and spacecraft. In reality, the rules of physics can precisely explain practically everything around you. Consider a smartphone: physics defines how electricity interacts with the device's numerous circuits. This information aids engineers in choosing the right materials and circuit architecture for the smartphone. Consider a GPS; physics outlines the link between an object's speed, its journey distance, and the time it takes to go that distance. Using a GPS gadget in your car uses these physics equations to figure out how prolonged it will take you to go from one place to another. Physics may substantially contribute to society via developments in new technologies resulting from theoretical discoveries. The quantum theory of contemporary physics was founded when German scientist Max Planck published his revolutionary investigation of the action of radiation on a "blackbody" material.

Planck established that energy might take on the qualities of physical substance in some instances via scientific experiments. According to classical physics theories, energy is only a continuous wave-like phenomenon that is unaffected by the properties of physical substance. According to Planck's hypothesis, radiant energy comprises particle-like components called "quanta." The hypothesis explained previously inexplicable natural phenomena like the behavior of heat in solids and the nature of atomic light absorption. Planck's study on blackbody radiation earned him the Nobel Prize in Physics in 1918.

Others, including Einstein, Bohr, Broglie, Schrodinger, and Paul Dirac, advanced Planck's theory and paved the way for the development of quantum mechanics—a mathematical application of quantum theory that sustain the energy can be both matter and a wave depending on certain variables. In contrast to classical mechanics, Quantum mechanics offers a probabilistic view of nature, which assumes that all exact features of things are calculable in principle. Modern physics is based on quantum mechanics and Einstein's theory of relativity.

References

https://www.grandinetti.org/quantum-theory-light

https://chem.libretexts.org/Bookshelves/Physical_and_Theoretical_Chemistry_Textbook_Maps/Supplemental_Modules_(Physical_and_Theoretical_Chemistry)/Quantum_Mechanics/02._Fundamental_Concepts_of_Quantum_Mechanics/Photons

https://www.scholastic.com/teachers/articles/teaching-content/origin-universe/

https://www.britannica.com/biography/Max-Planck

https://www.britannica.com/biography/Werner-Heisenberg

https://www.britannica.com/biography/Erwin-Schrodinger

https://www.livescience.com/32016-niels-bohr-atomic-theory.html

https://www.space.com/15524-albert-einstein.html

https://www.sciencealert.com/hawking-radiation

https://www.britannica.com/science/Hawking-radiation

https://www.britannica.com/biography/Kip-Thorne

https://www.nbcnews.com/mach/science/what-stephen-hawking-taught-us-about-black-holes-ncna857471

https://www.space.com/15421-black-holes-facts-formation-discovery-sdcmp.html

https://phys.org/news/2022-02-analysis-fundamentally-view-supermassive-black.html#:~:text=Located%20in%20the%20center%20of,that%20wander%20into%20their%20vicinity.

https://scitechdaily.com/supermassive-black-holes-on-a-collision-course-closest-pair-of-supermassive-black-holes-to-earth-ever-discovered/

https://www.sciencealert.com/10-mind-blowing-scientific-facts-about-black-holes

https://www.nasa.gov/audience/forstudents/k-4/stories/nasa-knows/what-is-a-black-hole-k4.html

https://physicsworld.com/a/what-you-need-to-know-before-investing-in-quantum-technology/

https://www.science.org/content/article/quantum-technology-comes-age

https://www.airbus.com/en/innovation/disruptive-concepts/quantum-technologies

https://www.tum.de/en/about-tum/news/quantum-technology

https://www.oxinst.com/applications/segments/quantum-technology

https://www.northropgrumman.com/what-we-do/disruptive-concepts-and-technologies-quantum-technology/

https://www.psa.gov.in/technology-frontiers/quantum-technologies/346

https://www.marconisociety.org/magazine/quantum-technology-past-present-future/

https://physicsworld.com/a/quantum-technology-why-the-future-is-already-on-its-way/

https://www.scmp.com/news/china/science/article/3161830/quantum-technology-how-it-works-applications-and-why-us-and

https://www.steadyrun.com/difference-classical-mechanics-quantum-mechanics

https://theconversation.com/seven-common-myths-about-quantum-physics-115029

https://byjus.com/chemistry/rutherfords-model-of-atoms-and-its-limitations/

http://abyss.uoregon.edu/~js/glossary/quantum_tunneling.html

https://www.dictionary.com/browse/quantum-tunneling

Printed in Great Britain
by Amazon

79966333R00068